My DOG IS DEAF

– but lives life to the full!

A practical guide for owners

Hubble & Hattie

The Hubble & Hattie imprint was launched in 2009, and is named in memory of two very special and much-loved Westies owned by Veloce's proprietors.
Since the first book, many more have been added to the list, all with the same underlying objective: to be of real benefit to the species they cover, at the same time promoting compassion, understanding and co-operation between all animals (including human ones!)
Hubble & Hattie is the home of a range of books that cover all-things animal, produced to the same high quality of content and presentation as our motoring books, and offering the same great value for money.

For Jack

First published in English in September 2011 by Veloce Publishing Limited, Veloce House, Parkway Farm Business Park, Middle Farm Way, Poundbury, Dorchester, Dorset, DT1 3AR, England. Fax 01305 250479/e-mail info@hubbleandhattie.com/web www.hubbleandhattie.com
ISBN: 978-1-845843-81-6 UPC: 6-36847-04381-0.
Readers with ideas for books about animals, or animal-related topics, are invited to write to the editorial director of Veloce Publishing at the above address.
British Library Cataloguing in Publication Data – A catalogue record for this book is available from the British Library. Typesetting, design and page make-up all by Veloce Publishing Ltd on Apple Mac.
Printed in India by Imprint Digital.

Contents

Foreword

'Disability doesn't define who you are' is a proverb which is often cited when discussing disability, and, for me, the quote sums up the idea behind this book. The word 'disabled' suggests that some sort of physical deficit is making a normal, happy life impossible. Even if this may appear the case at first glance, it most definitely is not. For it is not deafness that prevents a dog from leading a normal life, but his owner.

This book will give you the courage to accept your deaf dog for who he is. It will also help you to see that his disability is not a hindrance for him, and that his quality of life can be equally as good as those of his hearing peers.

The experiences described in this book are based on my co-existence with Chocolate, a deaf Dalmatian dog. The focus of this book is therefore on dogs with inherited deafness, although most of the tips and tricks detailed here can be readily applied to dogs with age- or disease-related deafness. Although there are certain factors that have to be taken into consideration, life with a deaf dog is basically the same as life with a dog that can hear. Both need training, and plenty of love and respect.

The advice on the following pages is not exhaustive, but is intended to provide food for thought. I hope you gain much satisfaction from using the ideas and motivation in this book for training and everyday life with your dog.

Introduction

On 23 May 2002, our dog, Chocolate, was born as one of nine puppies near Lübeck, in Germany. To search for our puppy, we decided to use a website dedicated to helping Dalmatians in need. Chocolate's breeder had advertised the puppy on this website (dalmatiner-in-not.de) because she thought that because of Chocolate's deafness, this website would be an especially good place for her.

Since we already had a Dalmatian dog named Jack, the decision to take on a new puppy was relatively quick, and Chocolate moved in with us. Despite her deafness, she developed like any other puppy: she attended puppy classes, and successfully completed a training course at dog school. Instead of responding to vocal commands, Chocolate responded to hand signals, learning new commands with obvious pleasure. Sometimes she got so excited she would perform her whole repertoire of tricks for just one treat! However,

Jack and Chocolate meet for the first time.

Chocolate knows her worth very well: people with no treats get no tricks – it's that simple! It's amazing how intelligent she is, because she seems to be able to count the treats that you hold in your hand. For three delicacies, she will hold up her paw exactly three times – if the hand is then empty, she turns her head away and refuses to 'work.' If someone was only pretending to hold a reward in their hand, she would never fall for it.

In this respect the difference between Jack and Chocolate was especially obvious. Even if Jack lifts his paw ten times and he doesn't get a treat, he remains loyal and happy to 'work.' He just continues to hope that a treat will appear!

Not so with Chocolate. Because she is deaf, she is more dependent on her observation of her environment – she always knows whether the reward really is something worth working for – and for Chocolate, treats are the only currency whenever effort is required.

Chocolate is a dog with a strong personality who knows exactly what she wants – and also how to get it. If her water bowl is empty, or the water doesn't appear good enough, she will pat my hand with a paw so that I notice she needs something fresh to drink. If she wants to be petted and cuddled, she jumps on the lap of her chosen person, despite the fact that she weighs at least 25 kilos (55lb)! She loves having a good snooze on a warm lap.

Chocolate is now eight years old and a wonderful dog. Her deafness doesn't impede her in the least; she lives a normal, happy, doggie life!

Baby Chocolate loved to sleep in a cardboard box!

Below & opposite: Chocolate and Jack play together, just like any other dogs.

Opposite: Baby Chocolate in the garden.

Chapter 1
Deafness: a death sentence?

For a long time, it was common practice to kill deaf dogs. The 'lucky' puppies were put to sleep humanely; others were dispatched by more brutal methods, usually drowning. Deaf dogs were regarded as worthless, because a breeder couldn't make any money from them: after all, they were 'imperfect' – and who would want damaged goods? Prejudice and ignorance had it that deaf dogs were aggressive, stupid, and had behavioural problems.

As animal welfare laws have improved, the number of dogs killed has, thankfully, continued to fall over the last few years. The experiences of many dog owners demonstrate that deaf dogs can lead a life that's just as satisfying and fulfilling as those of hearing dogs. Deafness in a dog is not a serious disease – it doesn't cause him pain or affect his life in any way that would make it not worth living.

The killing of deaf dogs is a crime under the Animal Welfare Act, and is punishable by imprisonment and a fine.[1] Of course, this doesn't mean that all deaf dogs are saved, as there are many unscrupulous individuals who breed dogs without regard to the health of the animals, in order to maximise profits. Popular breeds of dogs, such as Dalmatians, are especially affected by this practice. The breeder neglects their duty of care, meaning it is often the case that deafness is inherited by the puppies. Puppies that are disabled in this environment are still often killed, whilst others will

[1]Anyone convicted of cruelty to an animal, or not providing for its needs, may be banned from owning animals, fined up to £20,000, and/or sent to prison. Reference: www.defra.gov.uk/foodfarm/farmanimal/welfare/act/

be sold to unwitting buyers who are completely unaware that their puppy is deaf. The lack of veterinary care – because of the cost of it – inherent in breeding such as this may also mean that the deafness goes undetected.

If you are considering getting a dog, always seek out a reputable breeder, or go to an animal protection concern where you should receive honest and helpful advice.

How to find a reputable breeder

As a dog lover, you will want to invest some time in choosing the right breeder, to avoid falling victim to a dubious breeder. Puppies are often sold at market in areas such as eastern Europe, which attract dealers with rare breed dogs at bargain prices. These animals, however, are usually kept under poor hygienic conditions without benefit of animal welfare laws to protect them. Many puppies are shipped throughout Europe in their first weeks of life. Because they are separated from the mother far too soon, they are unable to develop normal social behaviours toward other dogs, let alone humans. Not surprisingly, these 'bargain dogs' usually have behavioural problems, and many are seriously ill and die within the first few months of life.

Even though this book aims to show that life can be just as great for deaf dogs as it is for hearing dogs, responsible breeding would greatly reduce the number of deaf dogs born. Unfortunately, some unscrupulous breeders use deaf dogs for breeding in the full knowledge that the disability will be carried on from generation to generation.

Therefore, it's a good idea to buy a puppy from a breeder who is a member of the Kennel Club, or someone who has come highly recommended by family or friends. On the British Kennel Club Association website, you can find information about breeders and breed clubs, and also breeding rules concerning animal welfare.

Of course, occasionally, reputable breeders may end up with a deaf puppy in the litter, and, when seeking the perfect owner for this puppy, they usually look for particularly trustworthy and caring people. If you acquire a deaf puppy from a reputable breeder, you can be sure that they will be on hand with plenty of help and advice for you.

Neutering/spaying

If your deaf dog is not neutered, seriously consider getting this done. Genetic deafness is inherited from generation to generation, and the castration of deaf animals can prevent this. The early neutering of a dog (for a bitch, this means

Use this checklist to distinguish responsible breeders from rogue breeders

- Beware of newspaper advertisements and internet services which offer various dog breeds at the same time. Reputable breeders focus on a maximum of two breeds at any one time

- A reputable breeder has nothing to hide, so insist on visiting the puppies in their first weeks of life one or more times. Not only will you get to know your future house-mate, it will also give you a clear picture of how the dgs are being housed and cared for, and you will see them with at least their mother

- The breeder should make a good impression on you, and genuinely care for the animals

- A good breeder should not trust just anyone with the puppies. He or she should vet you by asking about your personal and living circumstances

- Reputable breeders should not hand over their puppies until after the eighth week of life. by which time they should be vaccinated, wormed, and clearly marked (microchip or tattoo). The puppy should also have had a through health examination by a vet

- Trust your gut instinct: do the puppies seem healthy and happy to you? Do they run toward you playfully, and show interest in you and their surroundings?

- You should be suspicious if the mother can't be shown. Don't accept any excuses (she's at the vet, in the next room, at a friend's house, etc)

And one last thing: if you suspect someone of having a deaf dog put to sleep, or killed in any other way, do not hesitate to contact your local veterinary practice, or the RSPCA. The killing of a deaf dog violates British animal protection law, and should not go unpunished. Even an anonymous tip-off is better than not reporting it.

spaying before the first heat, or season) also reduces the risk of some diseases. In a bitch, there is less risk of a variety of tumours developing in the ovaries, cervix or uterus, or of her developing pyometra, a nasty disease of the uterus which, if left untreated, can be fatal. Neutering really makes a difference to health: for example, after the second heat, the probability of your dog developing mammary tumours is 26 per cent higher than if she had been spayed after her first heat, and inflammation of the uterus and ovaries can be prevented by early spaying. If you neuter (castrate) your male dog, you will reduce his changes of contracting testicular cancer and prostate disease.

The castration of male and female dogs has become a simple, routine procedure that will not limit your canine friend either physically or mentally. On the contrary, as already mentioned, early spaying and neutering brings significant health benefits. Of course, this is a big step, so thoroughly discuss the matter with your vet before making any decision.

Do the puppies seem happy and healthy?

Ethical dog breeding – interview with Daniela Gottmann

To give an insight to the problems of hereditary deafness in dogs, I did a short interview with Chocolate's breeder, Daniela Gottmann.

Daniela, born in 1966, is a doctor, and is married with four children. Since 1996, Dalmatians have been part of her life; her first dog, Cookie, was the grandmother of Chocolate. She currently lives with Amica (11 years old and Chocolate's mother), and Elvis (2 years old). Using the Kennel Club name 'Dalmatiner von der Palinger Heide,' between 1999 and 2008, Daniela has bred five litters, producing 44 puppies. Chocolate is the first and only deaf puppy.

Q *What are your personal observations regarding the treatment of deaf puppies/dogs in the various breeding associations? What has changed in recent years?*

A Previously, unwanted puppies were culled; that is, killed. With Dalmatians, these were the ones with flecked fur[2], blue eyes and deafness (blue-eyed dogs have a significantly higher risk of deafness). In the past, kennel clubs usually chose not to raise unwanted puppies, meaning they would put them to sleep. With 'defective' puppies, their 'shortcomings' were externally visible at birth. With deaf puppies, their defect cannot be detected until the second week of life, when they first hear. These would have been the animals that 'died suddenly' after a month or so.

In the breeding associations, various marketing strategies meant that only perfect puppies were wanted. Animal protection law outlawed this approach, and mandatory audiometric testing of all litters is now used to recognise deaf puppies. In my estimation, there have been changes made in relation to dealing with deaf puppies: earlier, before the introduction of audiometric testing in 1995, only a few deaf dogs slipped through the net by mistake, as the breeders may not have noticed their deafness. Thankfully, many breeders feel a sense of responsibility for their deaf puppies, and will pass on plenty of advice to a potential owner.

Q *What improvements would you like to see with respect to the treatment of deaf puppies by breeding associations?*

A I would like breeding associations to make it very clear to breeders that to kill an animal which is

[2]Dalmatians are born with a white coat, and their characteristic spots develop later on. Research shows that Dalmatians with large patches of colour present at birth have a lower incidence of deafness.
Reference: http://en.wikipedia.org/wiki/Dalmatian_(dog"http://en.wikipedia.org/wiki/Dalmatian_(dog)

not actually in pain and suffering is a violation of the Animal Welfare Act. In addition, I would like them to make it clear to breeders that it is possible they may produce 'imperfect' dogs, for whom they should accept complete responsibility.

Breeding associations should take a clear position when dealing with deaf puppies, in accordance with the Animal Welfare Act, which states that deaf puppies cannot be killed because they are not suffering. Compliance with this rule should be monitored and controlled.

Q *What is your personal experience with deaf dogs?*

A So far, I have only had experience with one puppy, Chocolate.

Q *What made you notice that Chocolate was different to the other puppies?*

A It was very early on, when the puppies began to hear, that I realised something was different about Chocolate, compared to her siblings. She was a very fit and agile puppy, but didn't respond to acoustic stimuli. Because I spent a lot of time observing the puppies, the difference was easy to spot. It was especially obvious in situations where she couldn't tell where her siblings were. If the little ones were sleeping deeply and there was a sudden loud noise, all heads went up in the air – except for one ... When I was feeding all the puppies, they were always close at hand when I called them, but Chocolate took much longer to come to me, at least until she realised that there was food. She learned naturally over time to rely on her siblings, so the delayed reaction was no longer so obvious.

Dalmatian breeders are usually aware of the potential problem of deafness in their puppies, and will do various tests to check their hearing.

Q *Was Chocolate treated differently after she was diagnosed as deaf?*

A Not really, because I knew weeks before the audiometric examination that she was deaf, so we trained her using more tactile and visual stimuli. For the first simple commands, it was important that she focused on us so she could see a hand signal. We encouraged this focus a lot and built on it during training. We also began using a kind of 'desensitisation' training: we took her out in lots of different situations to get her used to the world around her.

We noticed she didn't respond to sudden stimuli – unlike a puppy that can hear, she didn't even notice if someone approached her, although deaf dogs do seem to perceive sound vibrations better than a hearing dog, almost as if to compensate for the lack of hearing.

Q *What advice would you give to someone who was thinking of taking on a deaf dog?*

A The best thing to do is to acquire as much information as possible before purchasing a deaf dog, and be clear about the differences between a deaf dog and a dog who can hear. You will need to think about how you are going to train him (for example, with hand signals), and how to handle everyday situations (when to put him on a lead, etc). If possible, speak to someone who has owned a deaf dog – this will really help you because they will know what it's like to live with a deaf dog, and can give you plenty of advice.

Stages of development

Let's take a quick look at the first seven weeks in the development of healthy puppies.

In the first and second week, the puppies can nether see or hear, and, since the sense of smell is still not fully developed, the small puppies are guided primarily by touch, which they use to reach their mother's teats and judge the proximity of their siblings. Although the eyes and ears are operational in the third week, it's another couple of days before they are able to take in their environment properly. In the fourth to seventh week of life, all of their senses are developed and the puppies are bombarded with new impressions of the world around them. Congenital deafness cannot therefore be reliably detected until this age.

Interview with vet Alexander Heere

Alexander Heere practices as a vet near Hamburg. His specialist areas are audiometry, and preventive medicine.

Q *What causes deafness in dogs?*

A There are various causes. It can be an inherited genetic defect, or an acquired condition caused by injury to the tympanic membrane, tumour growth, or so-called ototoxic drugs, ie agents which affect the auditory nerve, damage the hearing, and can lead to deafness.

Q *Why does genetic deafness apply mainly to dogs with a white or predominantly white coat?*

A An allele (gene) which is responsible for white coat colour is also responsible for incorrect formation of the auditory nerve. So, the whiter a dog's hair is, the more likely there is to be a loss of hearing. Professor Distel at the

Hannover Veterinary School is currently conducting blood tests to examine this allele, in the hope of one day being able to use a simple blood test to determine which dogs possess this defective gene.[3]

Q *What are the different methods used to identify deafness in dogs?*

A Owners can test their dog for deafness by clapping their hands and watching the reaction of the dog (this does not rule out the possibility of single-sided deafness, however). Although the dog may react to acoustic stimuli, he may not be able to locate the sound. The most reliable method is audiometry[4], where nerve impulses are measured by EEG (electroencephalogram).

Q *At what age should a dog have an audiometric test?*

A Breeder clubs associated with the British Kennel Club Association advise carrying out the test from the eighth to twelfth week of age. Before that, it is, in my opinion, not very useful because the animals are not completely neurologically mature. Audiometry is also performed on older animals, if there is a suspicion of acquired deafness.

Q *What happens during the audiometric test?*

A The animal is sedated or placed under anaesthesia for the duration of the examination. The EEG has to be protected from as many outside influences as possible; even a small movement of the ear could distort the results.

The dog is then attached to each unit via three or four electrodes on the head, secured either with a plaster or with small needle electrodes. Then, a clicking sound is made via headphones or earplugs. If the dog is not deaf, the EEG will show corresponding fluctuations caused by stimulation of the nerves in the ear. The picture looks similar to an ECG, which is a graph showing peaks and troughs. The focus is on whether all appropriate peaks are visible, what the fluctuations are, and how great the distance between these peaks is.

Q *How accurate is audiometry?*

A The audiometry is very accurate,

[3]*Author's observation: Chocolate participated in the study of genetic clarification of canine congenital deafness by Professor Distel at the Hannover Veterinary School by donating a blood sample. For more information, visit: www.tiho-hannover.de*

[4]*For more information on audiometry, read:* New Handbook of Auditory Evoked Potentials *by James W Hall, Pearson, Boston, USA 2006*

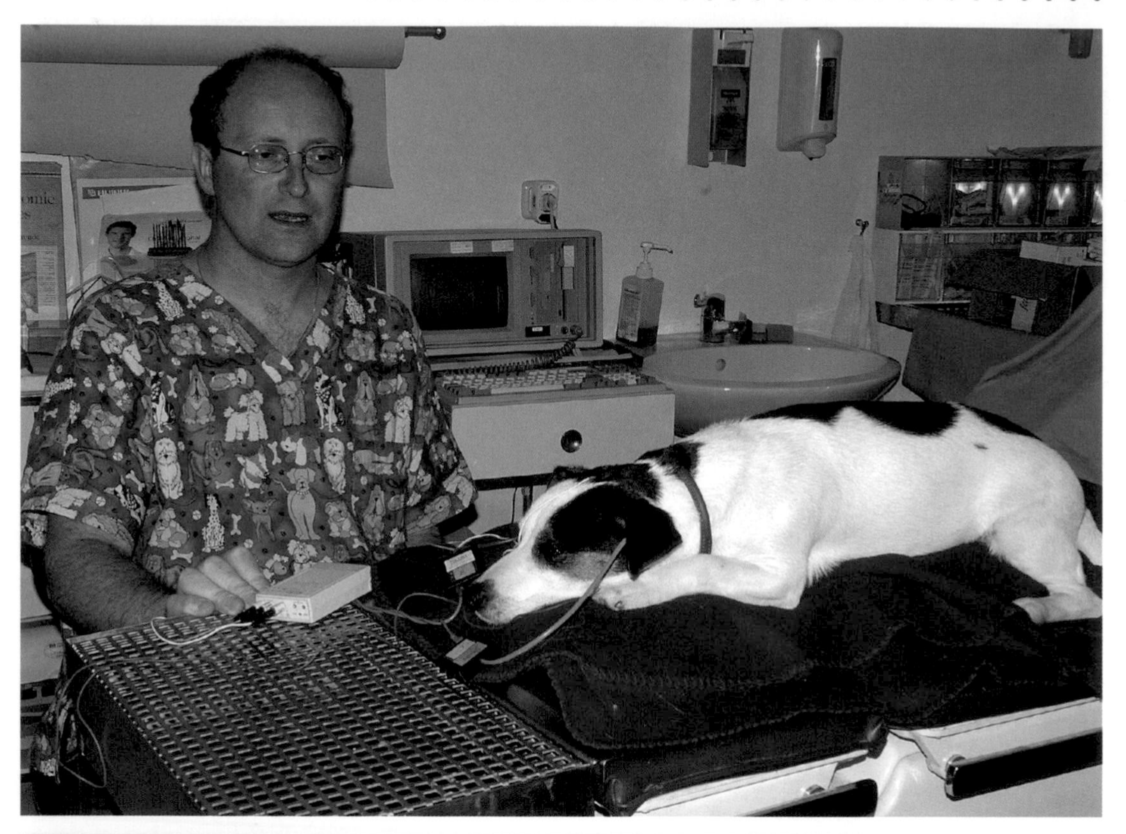

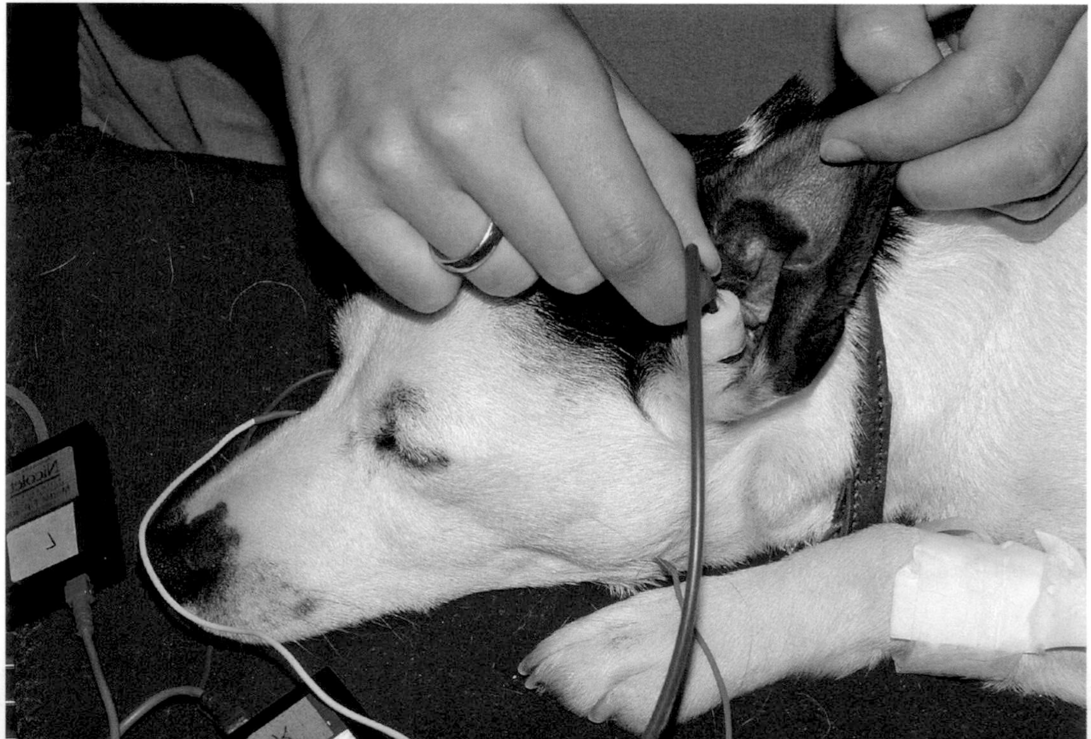

Top left: In order to obtain an accurate result, the patient must keep very still, and so is either sedated or anaesthetised.

Bottom left: Three to four electrodes are attached to the dog's head, to measure the amplitudes.

Above: Using a clicking noise over earphones, the vet stimulates the auditory nerve.

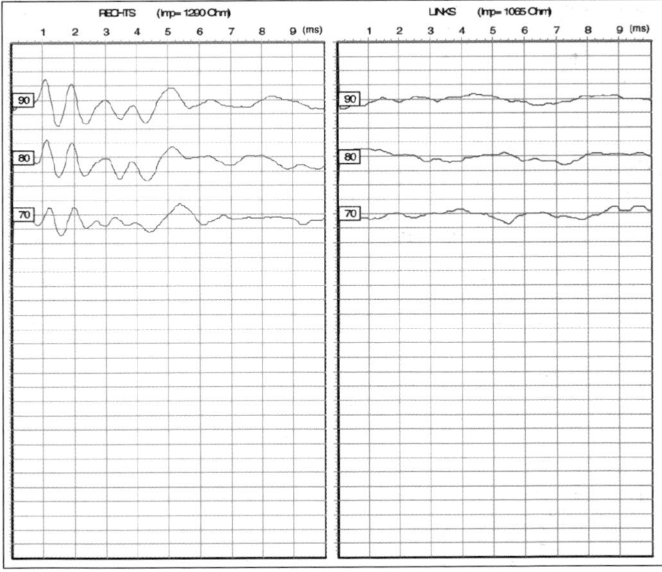

This audiometry result shows that the dog is deaf on one side only.

and allows us to determine whether a dog is deaf, and to what extent. It is possible, for example, to accurately locate the presence of a tumour which is suppressing the auditory pathway, or find out whether a patient is hard of hearing, has delayed hearing, or can only hear on one side.

DIY test

You can test your pup for deafness at home with a few very simple methods, though bear in mind that these test results alone are not conclusive. Even if your dog has responded to all your tests, he could still be deaf. Deaf puppies learn very quickly to make up for their shortcomings by using their other senses. Fine vibrations – tiny 'shocks' in the air – which go unnoticed by most people, are an essential part of a deaf dog's communication.

Therefore, it makes little sense to test your dog for deafness by pounding on the ground or clapping your hands in close proximity to him. Your dog will respond, even though he may not have actually heard the noise, because he is probably reacting to the vibrations, or what he can see.

Instead, try to test your dog using unmistakable noises which are not part of his usual environment. If your dog never reacts to other barking dogs which he cannot see or smell (on the TV, for example), the doorbell, your calling him from somewhere he can't see you, or the sound of food going into his bowl, this can be one of the many signs that your dog is deaf.

However, the only way to be absolutely certain is by taking him to have an audiometric test.

The canine ear

The canine ear is composed of external pinna (outside portion of the ear), and a skin-lined canal (the external ear canal) descending (vertical and horizontal portions) to the ear drum (the tympanic membrane). This portion comprises the external ear. Other compartments of the ear, located after the tympanic membrane, are called the middle ear and the inner ear. On the tympanic membrane, the malleus, anvil and stirrup pick up the vibrations that sound makes on the eardrum and pass them through to the inner ear. Sitting in the inner ear is the cochlea, which converts the vibrations to nerve impulses that transmit auditory signals to the auditory centres of the brain. The dog has 17 muscles with which to move his ear.

Although the size of the dog's ear varies depending on breed, all dogs can hear high frequencies 2-3 times better than humans, and low frequencies about as well as we do. Many dogs are able to move their

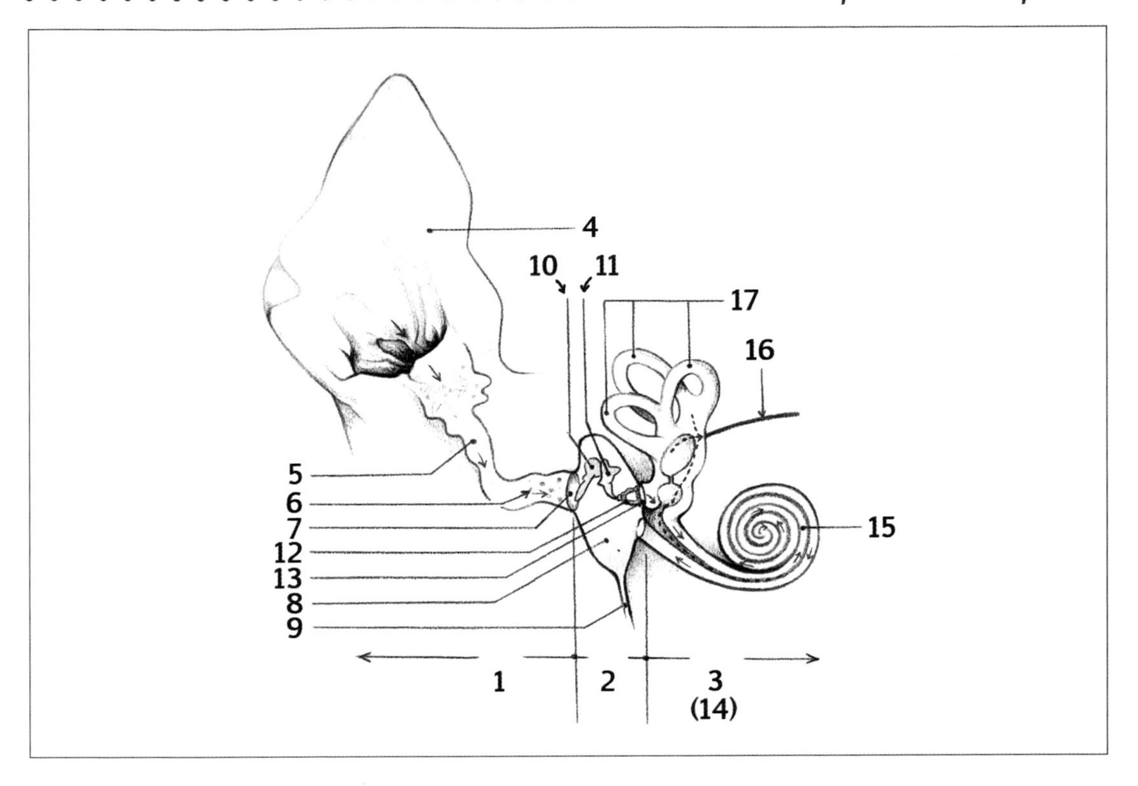

1 External ear
2 Middle ear
3 Inner ear
4 Ear muscle
5 Auditory canal
6 Sebaceous gland
7 Eardrum
8 Tympanic cavity
9 Eustachian tube
10 Hammer
11 Anvil
12 Stirrup
13 Oval window
14 Labyrinth
15 Cochlea
16 Auditory nerve
17 Semicircular canals

ears like an ear trumpet in order to catch sound, but for others, this function has been lost over the decades in breeding.

Interview with Dr Matthias Bruehl

Dr Matthias Brühl practices as a small animal vet in Koblenz, Germany, and has treated our dog, Chocolate. Although she is currently the only deaf dog among his patients, we feel his advice is always the best, and can't stress enough how important it is to find the right vet who can give you good advice on life with a deaf dog.

Q *What forms of deafness in dogs do you encounter most often in your practice?*

A Deafness in dogs is, generally, a rarity amongst my patients. Diseases such as meningitis, which, years ago, would have made dogs deaf or hearing-impaired, can now be treated much better and more quickly, so the number of animals that are deaf because of illness has declined greatly. More often, however, it is genetic deafness – though still rare – and age-related deafness that I have encountered in my practice.

Q *If my dog is diagnosed as deaf, does this mean his quality of life will be impaired?*

A My answer to this question is a definite 'no.' Firstly, the dog is not in pain, and doesn't realise he has a disability at all. He lacks a sense of hearing, but since he is not aware of this, he doesn't miss anything. Unlike we humans who have the associated difficulty of learning a language, dogs are not dependent on the spoken language. A deaf dog, therefore, has a completely normal quality of life, as long as the owner uses other modes of communication for mutual understanding. If the deaf dog lived in the wild, he probably would have significant disadvantages when it came to feeding and self-protection, but for our domestic dogs – who live in a close community with their people – there are no disadvantages at all.

Q *A common preconception is that deaf dogs have behavioural problems ...*

A I can confirm that this is a myth. The behaviour of a dog depends largely on the competence of his owner. If a dog is not socialised early on with other dogs and other people, and trained consistently, he will naturally develop behavioural problems. This is as true dogs who can hear as it is for deaf dogs.

Chocolate is a very quiet and good dog around the house.

Dog breeds that are at risk from hereditary deafness (as researched by the veterinary institute of the Louisiana State University)

Akita Inu
American Cocker Spaniel
American Eskimo Dog
American Staffordshire Terrier
American-Canadian White Shepherd
Australian Cattle Dog
Australian Shepherd
Beagle
Bichon Frise
Bobtail
Border Collie
Borzoi
Boston Terrier
Boxer
Bull Terrier
Bulldog
Canaan Dog
Cardigan Corgi
Catahoula Leopard Dog
Catalan Sheepdog
Cavalier King Charles Spaniel
Chihuahua
Chinese Crested Dog
Chow Chow
Collie
Coton de Tulear
Dachshund (White Tiger)
Dalmatian
Doberman Pinscher
Dogo Argentino
Dogo Canario
English Bulldog
English Cocker Spaniel
English Setter
English Springer Spaniel
Epagneul Breton
Fox Terrier
Foxhound
French Bulldog
German Shepherd
German Shorthair
Great Pyrenees mountain dog
Greyhound
Havanese
Ibizan Hound
Iceland Dog
Italian Greyhound
Jack/Parson Russell Terrier
Japanese Chin
Kuvasz
Labrador Retriever
Löwchen
Maltese
Mastiff
Miniature Pinscher
Miniature Poodle
Newfoundland
Norwegian Hound
Nova Scotia Duck Tolling Retriever
Papillon
Pointer
Puli
Rhodesian Ridgeback

continued over

- Rottweiler
- Samoyed
- Schnauzer
- Scottish Terrier
- Sealyham Terrier
- Shetland Sheepdog/Sheltie
- Shih Tzu
- Siberian Husky
- Soft Coated Wheaten Terrier
- St Bernard
- Staffordshire Bull Terrier
- Sussex Spaniel
- Tibetan Spaniel
- Tibetan Terrier
- Toy Poodle
- Walker American Foxhound
- West Highland White Terrier
- Whippet
- Yorkshire Terrier

As you can see, the list is (unfortunately) very long. Although you may feel very much alone with your deaf dog at first, please be assured that you definitely are not! There are more people with deaf dogs than you might think, all positive examples that the lives of deaf dogs can be as happy and carefree as those of their hearing peers.

Chapter 2
Living with a deaf dog

If you decide that you would like to take on a deaf dog, there are some questions you should ask yourself first.

- **What is your financial situation?**

When doing your calculations, you should not only take into account fixed costs such as food and – depending on the breed – the grooming required, but also costs that may occasionally arise unexpectedly; at the vet, for example

- **Are your living conditions dog-friendly?**

You will need enough space in your house, and somewhere the dog can call his own, a bed or basket. But you should also be able to offer something to your dog outside the home – ideally a fenced garden or a park nearby where there is no traffic if possible. A deaf dog needs to be able to romp around off the lead at full capacity, just like any other canine. Most behavioural problems result from insufficient exercise and/or stimulation

- **What about your family?**

Are all the people close to you in your daily life happy with your decision to look after a deaf dog? Is there anyone in your family who suffers from allergies or is afraid of dogs?

Deaf dogs tend to be very people-orientated and affectionate. Therefore, you should not leave him alone for too long. Does your lifestyle allow for this?

- **A deaf dog needs more attention**

If he is having great fun emptying out the bin or stealing something

As a puppy, Chocolate was just a curious as any hearing dog.

from the table, it is not enough to tell him off from the sofa across the room. He won't hear you. Therefore, more effort will be required to get his attention. Consider carefully whether you might find this too much to deal with at some point in the future. Dogs don't know what weekends, evenings, and holidays are!

- All of us have to leave the house on a regular basis, and sometimes your dog won't be able to go with you (for example, business trip, holiday, hospital, etc) This is fine, as long as we have someone we can trust absolutely with our four-legged friend. Anyone who is going to look after your dog has to be willing to learn hand signals in order to communicate with him, and must always use these in the same form as you do

- Your life could change overnight in one fell swoop – through marriage or divorce, birth of a child, relocation, unemployment, or illness So, if the worst happens, how will you take care of your dog?

● And the most important issue of all. Dogs can – depending on the breed – reach a good old age.

Chocolate loves running off the lead ...

... especially when she can do so with Jack.

No living creature should have to experience losing the safety and security of his beloved home and being deported to a shelter. Many millions of animals have had to go through this traumatic experience. Consider, therefore, very carefully, whether you are prepared for life with a deaf dog, with all that this entails.

I tend to compare the situation with preparing for life with a new-born baby. A deaf dog is no more self-sufficient than a child – he will need your help in all situations. And, unlike a child, who will eventually become independent, your dog will rely on your care for life; a very significant responsibility. If you already have dogs, consider whether you have enough on your plate with these. Don't assume your deaf dog will be guided by his new pack members. Initially, at least, plan to spend a lot of time rehearsing hand signals and building a strong bond between the two of you.

When your deaf dog comes to live with you

If you didn't actually decide to take on a deaf dog, and only retrospectively learned of his disability, you may be feeling sad, worried, or perhaps even angry. Because we were always aware that Chocolate is deaf, we never experienced these feelings, but I can assure you there is no need to be sad or angry, and owning a deaf dog is certainly nothing to be ashamed of. Your dog only knows life with his deafness – he is missing out on absolutely nothing, and he is happy with what he has, provided you accept him as he is. However, there are certain things you should pay particular attention to when you live with a deaf dog.

Things to consider as the owner of a deaf dog

It is very important to have a deaf dog microchipped or identified with a tattoo. Even if your pet is well-trained, he might get carried away if he sees a squirrel, when, all of a sudden, off he goes!

Microchipping is quick and painless; the chip is inserted under the skin (usually in the back of the neck) and the code easily read. If your dog is lost and a passer-by or the police take him to an animal shelter, they will have a microchip reader and be able to inform you immediately.

However, this will only work if you register the microchip. Many people think that when their vet puts the chip in the dog, it is automatically registered. But this is not always necessarily the case – you may need to fill in a form and send it off so that your dog is actually registered on the microchip system. Your vet should make this clear to you and the surgery will sometimes fill out the form for you – all you will need do is sign it.

Alternatively, you can get your dog marked with a tattoo. However, the disadvantage of this is that, over the years, the tattoo will fade and no longer be as legible. Discuss the pros and cons of tattooing and microchips with your vet.

In addition, you should always continue with the tried and tested means of identifying your dog, such as identity tags worn on the collar. On this tag, besides your contact details (but not your dog's name to deter thieves), should be the words 'I am deaf.' This way, the person who finds your dog will know why your dog doesn't respond to spoken commands, and won't think he is in shock, untrained, or injured.

This follows on nicely to the next point: always keep a particularly watchful eye on your deaf dog. As his carer, you are responsible for his conduct and behaviour, especially if you are walking along a public road. If your deaf dog suddenly charges off, you won't be able to simply call him back from a distance when he is looking away from you. Training and living with a deaf dog requires your presence far more than if you were living with a dog who can hear.

In many everyday situations, you will need to be inventive; for example, when devising new hand signals. This can be a lot of fun – and be aware that some people are going to stare. It is an unusual sight to see somebody communicating with their dog purely by hand signals, after all!

Some people will be fascinated and ask you questions, and some will be amused, but you may also receive offensive comments. In this instance, it's important you remain calm and indifferent to any negative reaction. Take on board the many positive reactions of those around you, and it will be easier to ignore a few negative remarks. Many people are simply ignorant of what deaf dogs are like, and judge them negatively as a result of this.

Tools to help you

In the beginning, when the bond between you and your dog is not yet strong, it may make sense to rely on certain aids when communicating with him. Many dogs love to chase a light, and a flashlight can be very effective when calling your dog to you in the dark (but only allow your dog off the lead in the dark in a fenced area).

Even if your dog is not facing you, you can use the torch to get his attention and draw it to you. Of course, you should never shine the light directly at his face or into his eyes, but on a point on the ground, so he can see it. Move the beam along the ground and then slowly toward you. Most dogs are curious and will follow the strange object on the ground. When he comes to you, reward your dog with touch and a treat immediately, so that he remembers why it's a good idea to do this (and he will improve more quickly this way).

Vibrating collars are available from pet stores. A handheld trigger makes the collar vibrate, attracting your four-legged friend's attention. Through patient training, you can get your dog to come to you when you activate the vibration collar.

Unlike the electric shock collar (which was finally banned in Britain last year by many animal welfare groups, including the Kennel Club), which is a completely unnecessary and cruel piece of equipment, the dog doesn't suffer any pain when wearing this collar.

Whether or not to use a vibrating collar is a matter of choice. The idea of being able to summon your dog any time at the press of a button is quite appealing, and, when used responsibly, the collar can be very effective in training your dog.

But technology has its limits.

Firstly, your dog must be introduced to the collar very carefully – he will find the vibration very strange at first, and will need time to acclimatise to it. Likewise, it will be a while before he learns what behaviour is expected of him when the collar is activated. The vibration collar is not a replacement for basic obedience training, but an aid to it.

Additionally, as most dogs love carefree play with other dogs, the resultant sensory overload (other dogs, people, thousands of different smells, and so on) could mean that the tingling in his neck can be ignored quite easily. Plus, if your dog always jumps in and out of water at every available opportunity, the technology could soon give up the ghost. If the collar fails in a dangerous situation, such as when your dog is around traffic, this could be fatal.

Therefore, never rely entirely on communication via a vibrating collar, although it may prove useful as an additional aid when training with hand signals.

Communication is everything – even without hearing

We, of course, communicate primarily through the spoken word, and, over thousands of years, the importance of facial expressions and gestures has faded. As deafness can affect us quite badly, a dog who can't hear seems to us to be a pitiful creature with a poor quality of life.

But when considering deafness in dogs, we shouldn't compare them with us, as the communication repertoire that he has at his disposal is much more sophisticated and versatile than ours.

Dogs have an extremely sensitive nose, which – depending on breed – has up to 200 million olfactory cells, whilst we have just five million. Not only is he able to detect subtle odours, he can very well distinguish the many different scents from each other. Roughly speaking, a dog's sense of smell is 40 times better than that of a human.

Smell also plays a very important role in communication between dogs, who recognise each other by their smell. When two or more dogs meet, they will sniff each other extensively. The information given and received – for example, age and gender – is recorded, and the dogs decide whether they will be friends or enemies. A dog's territory is eagerly marked with fragrances to signal to other dogs, 'I rule here, even if I only walk past this place once a day!'

There are countless examples of canine body language. The best known is probably the wagging tail. It can mean – depending on the situation and other body language signals – 'I'm happy, and I'm up for anything!' However, insecurity, aggression and anxiety are also indicated by tail movement. Baring the teeth, pulling back the ears, or a yawning wide grin (in some breeds such as Dalmatians); a dog can show many moods by his body language.

Your four-legged friend's posture is also significant when it comes to communication. A stiff, upright posture is intimidating and commands respect, whilst bowing down on the front paws with the bottom in the air is an invitation to play (the play bow).

So you can appreciate that hearing is only a small part of a dog's communication skills, and that body language is far more significant. Of course, this means that, as his owner, it is extremely important you become very familiar with the body language of dogs in general and your own in particular. Through careful observation, you will quickly be able to recognise if your dog is in a difficult or dangerous situation. Do a little research on canine body language (*Dog Speak: recognising and understanding behaviour*, published by Hubble and Hattie is recommended here), because this will enable you to better understand your dog.

Deafness in a dog is not the huge handicap it might at first appear to be: after all, who needs ears when they have a fine nose and sophisticated body language at their disposal?

Other species can also survive very well without hearing because their communication skills are not limited to language either.

Are deaf dogs dumb?

It might sound a little cynical, but

there are some advantages to living with a deaf dog.

As a child, I had a terrier who, typical of his breed, barked an awful lot. At the time it didn't bother me, but in hindsight, I realise it must have been very irritating for the neighbours in the flat next door. When the doorbell rang or someone stepped into the house, a cacophony of barking resounded from our apartment, which only stopped once the dog was worn out!

Also, our Dalmatian dog, Jack, was very alert and barked whenever the doorbell rang, or a car he recognised drove up the street. Chocolate usually ran to the door or gate when Jack started to bark and joined in – Jack lent her his ears, so to speak.

After Jack died, it became very quiet in our house. Chocolate can't hear the doorbell or any approaching cars, so no barking, and no more bad-tempered neighbours. For peaceful co-existence with your neighbours, a deaf dog definitely has advantages – especially if you live in an apartment building.

But the fact that a deaf dog doesn't respond to noises around him, doesn't mean that he won't use his voice, however. Chocolate knows, for example, how to use her voice to get what she wants. She has an amazingly large repertoire of various barks, howls, and other tones which are used precisely for each corresponding situation.

Chocolate doesn't like being alone, so, if she wants to go into the garden, she will run through the open patio door and wait on the patio, barking like mad (in a demanding tone) until a family member joins her – or we tell her to stop.

So, even though your dog is deaf, this won't mean that he is also dumb. Even though he can't hear approaching cars or visitors, you should learn how to quieten him with a hand signal so that he remains calm in certain situations.

Make it clear to your deaf dog from the outset that he should only continue to bark if you don't interrupt him. When strangers approach your house or your property, it may, for example, be handy if your dog barks. In all other situations where you want the barking to stop, put your left hand above his nose, and use your right hand to do the signal for 'No' (vigorously move your hand from side to side horizontally – see page 43).

If your dog barks or howls continuously, more than likely, this has nothing to do with his disability. Your dog is not barking out of malice or to annoy you, but could be doing it for one of the following reasons.

- Loneliness and the fear of abandonment

Deaf dogs, just like their hearing counterparts, need close and regular contact with humans, and are often afraid to be left alone. It is therefore important that you teach your dog early on to be alone for a short while and eventually he will learn that you always return to him

- Uncertainty and anxiety

Some dogs try to allay their uncertainty and fear with non-stop barking. This can happen, for example, if a dog is not properly socialised or is unfamiliar with his environment

- Boredom and feeling under-challenged

One of the most common reasons for persistent barking or howling is simple boredom. A deaf dog requires just as much exercise and stimulation as a dog who can hear. A walk around the block is not enough for these highly intelligent and sociable creatures. If your pet is not given enough to do by you, he may find something else to occupy him. Like barking!

- Master of manipulation

If your dog has learned that you rush immediately to him and do what he wants if he barks long enough he will use this advantage ruthlessly. Deaf dogs are not stupid – they know exactly what they want and how to get it! When you give him treats and stroke him, he will remember this and try it on until he gets attention from you again. Don't let him minipulate you in this way

- Canine dementia

As your dog ages, it may well be that he develops small quirks – exactly like us. Unfortunately, one of these quirks could be barking ...

- Blame the genes

Some dog breeds are especially disposed to barking; it's innate behaviour for them. This includes many breeds of Terrier. It is therefore particularly important, before buying a dog, to clearly and precisely check the characteristics of its breed, in order to avoid any nasty surprises later

- Setbacks

If you've rescued your dog from a shelter, it may well be that, in his earlier life, he had some negative experiences, such as violence and neglect, which can cause serious trauma. In situations that remind the dog of his bad experiences, he may bark out of fear and despair

If none of the above scenarios seems to fit your dog, consult an experienced dog trainer or dog psychologist. This is nothing to be ashamed of – on the contrary, you are being responsible and showing a duty of care. A dog psychologist can not only help determine the cause of the constant barking, he or she will also be able to offer help and advice on all sorts of issues. If you want to change your dog's habits, then this route is recommended, though bear in mind that change won't happen overnight. Patience and perseverance on your part will be required.

Separation anxiety

Of course, no dog, deaf or not, should ever have to spend the whole day at home alone, but for deaf dogs, it is especially important they are able to maintain a close relationship with their people, and be near them as often as possible. So, if you have a demanding job and can't take your dog to the office, you should ask yourself whether a deaf dog is right for you. That aside, of course, it's important that you are able leave the house for a few hours without your dog, but if he's not used to being away from you, you will need to gradually increase the time you spend away from him, starting with just a few minutes.

Leave your dog on his own for short periods of time, and keep him occupied with his favourite toy or chew. Leave the room and return after a few minutes, if he is happy and behaving. Act as if you have never been away by not making a big deal of your return.

Your dog will learn that it can be much more rewarding to pay attention to his bone or a toy than to follow you around all the time, and that there is absolutely no reason to panic, as you're not lost and will return. If this has worked well for a while, you can increase both time and distance gradually. This exercise will require patience – don't expect too much too soon from your pet.

The sooner you get him used to you being away for some time, the easier it will be for him. In fact, it is also possible to train an older dog in this way, though this will require a little more time and patience.

You may notice that your deaf dog is reluctant to let you out of his sight. Chocolate is always very keen to ensure that a family member is close by, and is happiest when we are all in the same room as her. As soon as someone leaves the room, she tries to follow to find out where they are going. You may have similar experiences. If possible, let your dog know when you are leaving the room and give him the opportunity to follow you.[5]

[5]*This tip contradicts what experts advise in order to avoid separation anxiety in dogs who can hear, but, in our experience, deaf dogs need to be dealt with differently. Of course, this depends on your dog's character – as with every training tip.*

If your dog hasn't noticed that you have gone, when he does, he may feel anxious or even panic, and search the house for you. Get your dog's attention before you walk away from him. If he is not looking at you or is asleep, just pet him or otherwise draw his attention. You don't need to signal to him; it is sufficient that he sees where you have gone, if he follows you, that's up to him. In situations where he can't accompany you, give him the command for 'Stay!' (see page 43). Your dog will soon learn to distinguish when he can and can't go with you, and will realise that you never leave him permanently. Initially, he will probably follow you quite a lot, until he has understood

Chocolate feels safe and secure in her bed; it's 'her' place.

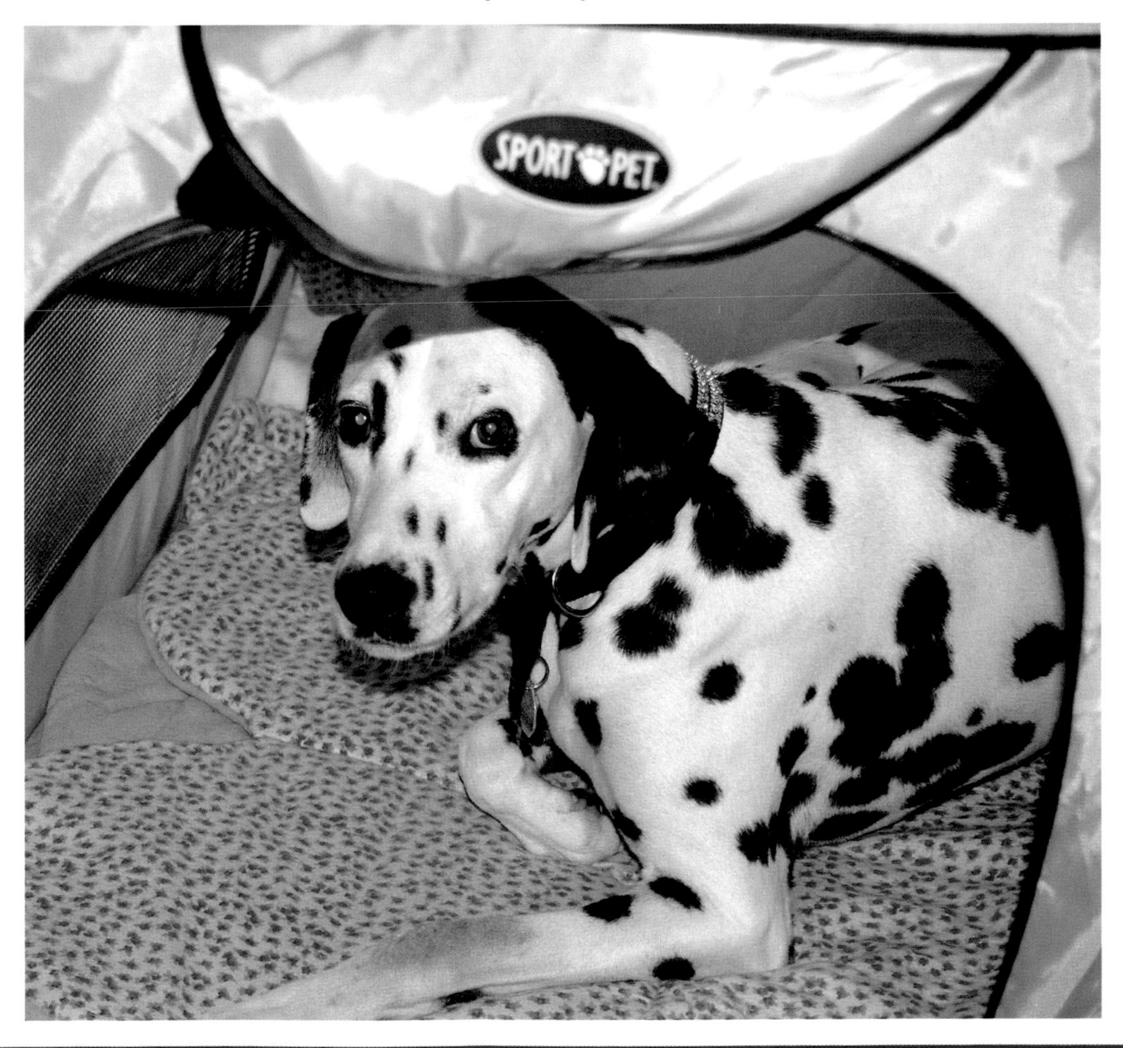

this and realises that he's more comfortable in his basket, safe in the knowledge that you will always return.

Deaf dogs and children

It's no secret that animals, especially dogs, can have a positive influence on a child's development, as numerous studies prove. Dogs promote both social skills and self-confidence in young people and, at the same time, teach them how to have a responsible and respectful relationship with animals. Children with behavioural problems, or those who are very unwell, can develop a more positive attitude by spending time with a much-loved pet.

The fact that your dog is deaf is absolutely no reason for him not to spend time with children. Because of his strong dependency on the people around him, he will be glad of the extra attention and care.

However, there are a few rules that you should follow when dog and child meet.

- Making introductions

Even if your dog has been raised with kids and they love him more than anything, this doesn't mean that other children can drape themselves around his neck without him batting an eyelid. Just as we are not keen on close contact with total strangers, neither are our dogs.

Ask the children to first of all introduce themselves to your dog by holding out their hand for him to sniff. If your dog is happy with this, allow the children to stroke him. This applies not only to children but adults also; your deaf dog is already busy scanning his environment with his eyes and nose, and may not notice if someone approaches him from behind. If he then becomes alarmed, it's possible he could snap out of fear, so letting strangers touch your dog before they have introduced themselves to him is not a good idea

- Dogs deserve respect

No one enjoys being angry or frustrated. Dogs are very sensitive in certain anatomical places, such as the face and paws, and shouldn't be touched here under certain circumstances.

Your four-legged friend should have his own bed or blanket, and your children should learn not to disturb him when he heads for this refuge.

His toys should also belong to him alone

- Eat in peace

Dogs have descended from the wolf, so it is natural that they should want to defend their food and prey against potential competitors, and children should leave the dog alone when he's eating.

In the worst case scenario,

the dog could snap at them as a defensive gesture. It could be that, when eating, your dog doesn't realise that someone has approached from behind (because he can't hear them), and is startled.

Also, playing right after eating is taboo: large dogs can suffer from a life-threatening bloat if they rush around too much after eating

- Expect the unexpected

Even if your dog is the sweetest, most reliable dog in the world, he can still behave unpredictably. Therefore, always be prepared for the unexpected, and keep a watchful eye on the overall situation.

Teach this way of thinking to children who have contact with your dog

- Keep calm

Ensure children understand that they must try and keep calm should your dog suddenly snap. If they panic and show fear, the dog may also panic and bite again

- Never run away from dogs

This follows on from the last point.

Many dogs chase their moving prey. If children are afraid and run away from a dog, this may arouse his 'prey drive' and encourage him to chase them

Do not leave children unattended with your dog, even if he is very well trained and fond of children. Even sensible, animal-loving children may accidentally inflict pain, or startle your dog in some way.

Dogs and children can make a dream-team, as long as both follow a few rules of play. Josie and Immie are great friends!

Chapter 3
Training

Some good training will help life with your deaf dog run smoothly, and avoid the possiblity of him getting into all sorts of mischief in the house.

To train your dog properly, you must first become familiar with his natural social behaviour. Despite all the years of domestication, his natural behaviour is strongly reminiscent of that of his ancestor, the wolf. For example, within the pack, a clear hierarchy exists, which, unlike humans, does not take into account gender.

Today, the 'pack' of the domestic dog is usually the family he lives with. Dogs, in general, operate differently to humans, and deaf dogs are in a special group again because they lack a particular sense, for which they compensate in other ways. From day one, you won't be doing your deaf dog any favours if you let him get away with bad behaviour out of pity. Remember: your dog can live a normal life; he is doing great – he doesn't need pity! If you let your dog get away with murder because he is a "poor deaf little poppet," then he won't take you seriously or respect you! An owner who doesn't deal with his dog consistently and logically will more than likely be ignored, so treat your deaf dog with the same rigour and discipline – and just as much love and affection – as a dog who can hear. Only in this way can you develop a healthy dog-owner relationship.

It's essential that you identify the boundaries right from the start – and you will need to be consistent. Use clear rules and commands. There are many things we do that have the potential to confuse our dogs; for example, exceptions are an alien concept to them. If you sometimes let your dog sleep on your bed, but other times not, he will not understand why, and it will demonstrate to him that you are not a good master because you keep changing your mind! Dogs need a clear set of rules, so decide what he will and won't be allowed to do long before you bring him home to live with you.

The methods you can use to train a deaf dog aren't all that different to those that are successfully applied in the training of hearing dogs, although certain things won't work. For example, you won't be able to call your dog whilst sitting comfortably on the sofa if he cannot see you. The most important thing when training a deaf dog is to ensure that you always keep things interesting for him. Your training will only work if you encourage him with treats, petting, play and fun so that he wants to stay close by you and actively seeks out your company. And this, of course, is the biggest difference with living with a deaf dog: he needs to be in close proximity to you far more than a dog who can hear. Without you being actively involved, training will not work.

When training a deaf dog, it can also be very useful if you are already living with a second, hearing (and well-behaved) dog, as he should quickly guide his new companion and demonstrate the desired behaviour to him. A well-behaved dog is not a replacement for training, but he can have a positive effect on your deaf dog.

Another big difference in the training of a deaf dog is, without doubt, the method of communication between you and your pet.

Talk to the hand – commands using hand signals

When you communicate with your dog using hand signals, it will probably seem strange at first, and some commands are also likely to attract the attention of others, who will be very curious.

It helps if you speak as well as making the hand signal, despite the deafness, as your gestures may be more convincing this way – after all, it's not uncommon to accompany words with gestures. Deaf dogs are usually very attentive, and pay very close attention to the facial expressions of people. If you are telling off your dog, for example, you can communicate this via hand signals, as well as actually saying 'No!' Your facial expressions will put more emphasis on this command in order to make it clear to your four-legged friend that his behaviour was not satisfactory. Of course, with some practice, you will be able to match the hand signals with the appropriate facial expressions.

Since we humans communicate primarily through language, however, it will probably help you to express additional commands out loud.

Incidentally, the zoologist and dog expert, Dr Patricia McConnell, has discovered that dogs generally learn hand signals easier and better than spoken commands![6]

The following commands are, of course, only a few examples of what you can teach your dog. 'Come' 'Sit!' 'Heel!' and 'No!' are all used as part of basic obedience, and should help everyday life run a bit more smoothly for you and your dog.

There are many ways to teach the most important commands to your dog, and it is vital that both of you are comfortable with how you do this. Be patient and don't despair if your dog doesn't seem to understand what you want from him the first time round. Every dog has his own pace, and his own talents and weaknesses. If you repeat your exercises consistently and persistently, practically every dog can be trained to respond to hand signals.

You can use any kind of hand signals you want, as long as you are expressive and clear with your signals, so that they are not confused with other visual signs. In addition, don't change the meanings of visual signs once they have been learnt. The hand signal he has learnt for 'Sit!' must stay the same, otherwise your dog will become confused.

For ideas for different hand signals, it may be a good idea to do read other books about sign language for dogs.

Hello! I'm talking to you!

Communicating with your hands only is not possible if your dog is not looking at you. Say good-bye to the idea that your dog will automatically look at you when you give him a command, but there are several ways that you can get your dog's attention:

- If he can see you out of the corner of his eye, signal to him that you want his attention. Then he will look at you and you can communicate with him
- Wink at him if he can see you. Then he will be looking at you and you can begin your communication
- Walk by or around him with a heavy tread so that he feels your footfalls
- Turn the light on/off
- Clap your hands in his vicinity
- Nudge him gently
- Close/open the door; the draught will make him aware of you

Most importantly of all, always have treats in your pockets and reward him in an almost over-the-

[6] *Suggested further reading:* The Other End of the Lead, *by Patricia B McConnell, Kynos 2004.*

top way when he looks at you. Even if you don't have treats, ensure this response is always praised.

Praise

During all of the exercises – and, of course, for good behaviour in everyday life – you should praise your dog immediately and directly, so that he associates the praise with his behaviour. A lot of dogs are food-orientated, so it's worth using dog biscuits and treats in this case to reinforce the training. Kind words (your dog will immediately notice the loving expression on your face) and affection are also important rewards for your dog, particularly if food does not do the trick.

There are two versions of a signal for 'well done;' here's the first …

… and this is the second.

I'm so glad to see you!

Wagging the tail and wanting to be near you is perhaps the most important body language of all, and says quite clearly: 'Glad you're here; I'm delighted to see you.' Deaf dogs become extremely attached to their people, which can actually be a big advantage, as you won't have to do much to get your dog to come to you. Always have a few treats in your pocket (at least at the start of training) and be ready to reward correct behaviour.

Whenever your dog looks at you, or is already on his way over to you, use a hand signal (eg wave) to encourage him to come to you, with a huge smile on your face and a delicious snack as a reward in your hand. You can use this hand signal, for example, when you get home and your dog happily welcomes you, or if you have left the room without your dog and he follows.

Check that he's always looking out for you and sticking close by (essential if you want to walk him without a lead, although only if he is happy to stay close by, and is always focused on you; see next section).

Come here!

For your deaf dog, regular exercise without a lead – where he can really race around properly – is very important. There may be an occasion when you need your dog to come to you immediately, so use the hand signal for 'Come!' right from the very start. Hold your arms out in front of you, palms upward, bend your arms at the elbow and raise them over your shoulders: this is the signal for 'Come!' Once your dog is with you, reward him with a treat and praise, and he will soon learn that it's a good idea to come when called.

Chocolate knows it's always worth coming to her owner when he calls!

Sit!

This command is extremely practical and a vital part of your training repertoire. You can practice it when on a walk with your dog by asking him to sit whilst another dog passes by. This command is not difficult to learn but does require some practice.

Kneel in front of your dog – who should be standing with a wall, or some other solid structure, directly behind him – and hold a treat in your left hand. Hold the treat over your dog's head, while making the hand signal for 'Sit!' with your right hand. Point the index finger of your left hand, and gradually take the hand back over his head toward the wall. In order to see and follow the treat with his eyes, your dog will look up and then sit down to continue watching your hand. As soon as he does, reward him and do the hand signal in front of him again. After a few repetitions, he will soon understand what you want him to do. The pointing finger means 'Sit!'

The raised, pointed finger means 'Sit!'

Down!

Once your dog has mastered the 'Sit!' command, it's time to go a step further. The command 'Down!' is, amongst other things, a practical command for occasions where you are out and about with your dog. Wherever you might be, with this hand signal you can let your dog know that everything is under control.

Begin the lesson by first asking your dog to sit. Show him that you have a treat held between the thumb and palm of one hand. Run this hand over your dog's nose to the ground. Then open your hand, keeping it parallel to the ground, palm facing downward. He will follow your hand and lie down, whereupon give him the treat. This hand signal may take longer to teach but remember that practice makes perfect.

Heel!

The 'Heel' command is always useful if you go out with your dog, especially in traffic or anywhere where there are crowds of people. This exercise isn't easy for your dog, and it may take a while until he understands what you want from him.

For this command, your dog should be walking at your pace, close to your left leg so that you have him under control. I'm sure he would much rather run ahead and sniff around, so be patient with him. It's best to attempt this exercise after your dog has been allowed to investigate the nearest lamppost! Ask your dog to sit on your left,

If Chocolate's owner places his palm parallel to the ground, she knows it's time to lie down.

If we want Chocolate to heel, we tap the palm of the hand on the thigh.

close to your foot. Keep the lead short and signal to your dog by tapping with the palm of your hand on your left leg, showing him that you have now begun. The tapping is the hand signal for 'heel.' Walk a few small steps and ask him to sit again. Praise him if he remains close to you during these first few steps. You shouldn't feel any tugging on the lead during the command, and cover very, very short distances initially, otherwise you run the risk of your companion being distracted by passers-by, noise, or other animals. If he manages small distances without too much trouble, you can gradually extend these.

Stay!

The 'Stay!' command is also useful for everyday situations.

Begin the exercise by asking him to sit or lie down, and then combine that command with 'stay!' the hand signal for which is to bend your arm and raise your hand, palm outward. Back away slowly one or two steps. Praise him profusely if he remains where he is. Let him stay there for a while (a short time at first). Your ultimate goal is, of course, that your dog remains where he is, even when he can no longer see you.

The two parts of the sign for 'stay close.'

No!

Dogs can get themselves into all sorts of trouble. Sometimes it's the furniture that suffers, or a pair of shoes lying on the floor, or food that has been left unattended. When this happens, it's very useful to be able to signal to your dog that this behaviour is not acceptable, and the following hand signal will help you deal with bad behaviour.

Bend the arm at the elbow so that the lower arm is a right angles to your body, palm facing downward. Move the lower arm from side to side, keeping the palm facing down. Whenever your dog is doing something he should not, immediately get his attention and show him the 'No!' hand signal. Combine the hand signal with a firm oral command of 'No!' Your dog can't hear this, of course, but he will pay attention to your facial expression, and notice immediately that you are not smiling at him.

Whenever your dog has behaved improperly, and you have used the 'No!' command, ask him to follow another command, such as 'Sit!' say, and praise/reward him for this, thereby ending the interaction on a positive note.

The sign for 'No!'

The two parts of the sign for 'Find!'

Bed!

It can be quite useful if you can signal to your dog that he should go to his bed. It's fairly easy to incorporate this into your daily routine, as your dog may naturally go to his bed for a time during the day.

When you see your dog lay down in his bed, give him the sign for 'Bed:' rubbing the palms of your hands against each other. Then give him a reward in the form of a treat. Your dog will very quickly come to realise that there's a reward on offer when he goes to his bed. You need, as always, just a little patience, and always reward him with a treat when he does as you ask.

By the way, even deaf dogs don't want to 'hear' every now and then. Chocolate will turn her head when she doesn't want to obey a command, as if she's saying something along the lines of: "I'm sorry; are you talking to me? I didn't see you there!"

Never be afraid to take advantage of professional help. Some dogs just need more training than others when it comes to learning commands. A dog school or a canine behaviourist will be happy to provide advice and assistance.

If Chocolate doesn't want to go to bed, she will simply ignore the hand signal by turning her head away!

Clicker training

Clicker training is a way to condition a dog to carry out desired behaviour: when the dog does as he is asked, the clicker is clicked and (at least initially) a treat is given, thereby reinforcing the behaviour and associating it with something nice.

Of course, a clicker is not a great idea for a deaf dog since he can't hear it, though it is still possible to take advantage of the clicker method by using an LED light or laser pointer, for example. If your dog has mastered an exercise well, show him the light as confirmation of this. When on the move, this method is not always practical, although there are solutions, such as a small laser pointer or LED light that you can clip onto your key fob, but you can also use the 'Well done!' signal (thumbs-up). Unlike a torch, you can't forget your thumb and leave it at home!

Let sleeping dogs lie...?

As the saying goes, you shouldn't wake a sleeping dog, and there

Rewards are guaranteed to encourage results. A 'thumbs-up' signal is a good alternative to the clicker.

is some truth in this adage. Dogs sleep a lot, and for a puppy, plenty of sleep is essential for physical and mental development, so ensure that your dog gets enough rest.

Having said that, you should accustom your dog to being woken by you from an early age. Then, if you have an urgent appointment, say, you won't have to wait until your dog wakes before you can go.

But how to go about this? From time to time when your dog is sleeping, very gently nudge him, blow softly in his ear, or tap the floor close by with your foot. You can also hold your hand in front of his nose and wake him with your scent. The important thing, of course, is that you wake him gently and not tear him from his sleep. Once he is awake and attentive, praise him with attention and a treat, so that he links his awakening with something positive. A smile from you will contribute to the good mood.

When Chocolate is asleep, it's important to wake her gently and carefully.

Also, when he is awake, you should 'desensitise' your dog by occasionally softly touching him when he has his back to you and can't see you, praising him as you do so. This will gradually decrease his fear, should he be awakened or touched unexpectedly. Ensure, however, that strangers do not wake him from his sleep or touch him as a surprise, and this will avoid any unnecessary problems.

Urkunde

~~Herr~~ / Frau

Willms

hat vom 4.4.2003 bis 6.6.2003

mit Hund Chokolate

an dem

"HUNDEERZIEHUNGSKURS"

- Grundkurs -

in dem Verein für

Deutsche Schäferhunde (SV) e. V.

Ortsgruppe Rübenach

teilgenommen.

Dozing on the terrace in the sun is Chocolate's favourite pastime.

Hiermit bestätigen wir dem Team

Frau Willms mit Chocolate

die Teilnahme an dem 10-stündigen

Welpenkurs

der OG Rübenach vom 16.08. – 18.10.02

Wir wünschen weiterhin noch viel Spass und Erfolg bei der Hundeerziehung!

(Unterschrift)

Chocolate's obedience and puppy school diploma certificates: deaf, yes; stupid, no!

Chapter 4
Out and about with your deaf dog

When out walking with your dog, you're bound to meet plenty of people who are animal lovers, but, even if your dog is very people-orientated, friendly, and well trained, think twice about allowing him to

48

be touched and stroked by all and sundry. Not every dog welcomes this kind of attention, but, if you are happy that he is, ask that the individual first introduces themself by letting your dog sniff their hand, and doesn't make any sudden movements or take him by surprise.

In traffic

The most important rule when out and about in traffic is this: never let your dog run around off the lead (in England, it is against the law not to have a dog on a lead on a designated road). No matter how well trained your dog is, and how usually reliable at responding to your hand signals, many unpredictable situations can occur when your dog isn't one hundred per cent under your control. Remember that your four-legged friend can't hear vehicles, and could suddenly run into the road, with potentially disastrous consequences for him and others. Even if your dog is good on the pavement, danger could be lurking round the corner: watch out for junctions where traffic could pull out without seeing you and your dog. Not everybody that you will meet is a fan of dogs, and some people are even afraid of them. Small children and elderly folk could fall if your dog jumps up at them.

On the bus and train

Even if your dog can't hear noise whilst using public transport, the many different smells and vibrations might initially frighten and confuse him. As you approach the bus or train, walk your dog very slowly toward it, and only travel short distances at first, rewarding him for his bravery and calm behaviour with petting and treats. During the journey it's best to ask him to sit, and make and receive as little contact with other passengers as possible. He will soon realise that the movement of the vehicle won't hurt him, and the combination of the journey and reward will hopefully mean that he comes to find this a positive experience.

Daniela Gottman with her dogs, Amica (left), and Elvis. Amica is Chocolate's mother.

Chapter 5
The prey instinct

There's another important reason why your dog must learn to consistently obey your commands if you want to walk him off the lead. Many of the popular dog breeds were originally bred for hunting or herding, which includes those that are at risk of genetic deafness.

Because of their wolf ancestors, all canines have an innate hunting instinct, meaning that a walk in the countryside could quickly become a nightmare. It's not easy to control the hunting instinct of your dog, but it is possible through basic obedience.

With some dogs, the hunting instinct is so strong they can never be allowed to walk off the lead. If your dog stalks a wild animal, this is officially classified as poaching, and your dog could be in danger from a gamekeeper attempting to protect the wildlife. In addition, as the dog's owner, you could end up with a hefty fine. Once your dog has gone through the hedge and has left your field of vision to chase a deer, hare, or pheasant, you have no chance to recall him. Even if he doesn't kill or injure a wild animal, he might get lost and not be able to make his way back to you, and could run onto a road and get run over.

Even if your dog has little or no hunting instinct, you should put him on a lead in the spring and summer months, when there are more wild animals about. In a set-to with a wild animal, there's the potential for your dog to be injured, or become infected by diseases, for example, worms, from the faeces of wild animals.

So you can appreciate that 'Heel!,' 'Come!,' and 'No!' are essential commands which your pet must learn and obey without exception.

Incidentally, many deaf dog behavioural problems are caused by the owner. Anyone who exercises his dog enough, mentally challenges him and allows him to spend time with his own kind, will usually have a well-balanced, happy dog – and that goes for deaf dogs as well!

Be confident that your dog will do as you ask and "Stay!"; even when you encounter wildlife such as hare and deer.

Chapter 6

Oh, the poor dog!

Dealing with prejudice

First of all, it's important to note that prejudice is not always generated by bad intentions. Most people simply don't know any better. A typical example is a visit to the vet with Chocolate. You can probably imagine the scenario: the person next to you is talking non-stop to his pet while they wait for their appointment. He suddenly turns to your dog: "And who's this then?"

You feel compelled to tell the man: "She's deaf, she only responds to hand signals." Then comes the pitying look. "Oh, the poor animal." Although he means this kindly, it is still a form of prejudice. Most people are convinced that deafness means social exclusion and a detrimental affect on quality of life. They transfer their own fear of disability to the dog, but, as already described, we can't possibly judge how a deaf dog feels using human standards.

So how do you deal with this inappropriate and unnecessary pity? One way, of course, is to nod and ignore the remark. However, remember that these attitudes are the exact same ones that lead to many deaf dogs being put down. The best evidence to the contrary is sitting right in front of you. Your dog is missing out on absolutely nothing except his hearing, and he doesn't need pity – you could try politely explaining all of this to your acquaintance.

"Do I look handicapped to you?"

Chapter 7
Fun and games

Don't assume because your dog is deaf that he can't take part in classes at a dog school. Although, unfortunately, some schools and clubs are reluctant to work with deaf dogs, or even completely refuse to take them, don't be put off; choose a club or instructor who takes your dog as seriously as they would those who can hear. In fact, many dog schools actually prefer to work with hand signals, or use them alongside verbal commands.

If you have a puppy, it's a great idea to register him for puppy classes where he will learn how to interact with other dogs, and develop normal social behaviour. The sooner a young puppy is socialised with other dogs, the more balanced and peaceful walks and meetings with other dogs will be for you both.

For most dogs, a few laps around the block every day is not sufficiently demanding physically and mentally. Dogs love to have a job to do – on the training field he can let off steam, and the teamwork required will strengthen the bond between him and you.

Activities

Once your dog masters the most important basic obedience commands and can follow these perfectly, you can take him to all sorts of activities. For example:

Companion dog certification

This testifies that your dog knows how to behave well in public: when meeting with pedestrians, cyclists, or other dogs, he remains calm and follows your instructions. You, as a dog owner, show your expertise when it comes to being in control of your dog. Companion dog certification is the gateway to many more exams you can do with your dog.

A FEW WORDS ABOUT COMPETITIONS AND EXAMS

Again and again, deaf dog owners are told that they cannot take part in the Kennel Club Good Citizen Dog Scheme and such like, and this is simply not true. Deaf dogs cannot be excluded from a tournament if the examination regulations state that hand signals can be used, so check out the examination regulations to establish the situation. If visual signals are permitted or not expressly excluded, you could contact the examiner, let him or her know that you want to participate with a deaf dog, and tell them what hand signals you use. If you discuss these things in advance of the exam, in most cases, it shouldn't be a problem.

Some individuals involved in dog sports believe that dogs should be perfect, and are obsessed with achieving this. Deaf dogs don't fit into this picture of perfection, of course, but don't be put off if you come across such a person, simply find another dog sports club, whose aim is that working with dogs should be fun.

KEEPING FIT WITH AGILITY TRAINING

Once your dog knows the most important basic commands, you can make his day by taking him to agility classes. This activity involves

Leading a dog through an agility course using hand signals.

your dog learning to go around a course of various obstacles, which could include slalom routes, seesaws, tunnels, and trenches.

It's not so much about how fast he can complete the course, but how smoothly the various different sections are dealt with. Since agility is done off the lead, you will have to guide your dog using only hand signals, so it's obviously important that you are a good team, and that your dog knows the basic commands perfectly.

You will also need to be fairly fit: because you are controlling your dog with hand signals, you'll need to remain in his field of vision, which means running with him. This can be quite tiring if your dog is really speedy, but a lot of fun, too!

Dancing with your dog – Dog-dancing

Dog-dancing is a fairly new concept which involves moves that your dog can do such as standing up on his hind legs and spinning around, and elements of a varied choreography routine. Again, you direct your dog just by hand signals, so very good basic obedience is also vital when doing this activity.

When done to music, the impression given is that you are dancing with your dog, but your deaf dog won't have a problem, as even hearing dogs aren't really following the music as we would. However, your choreography will look its best if it corresponds with the music.

Retrieving with a difference – Frisbee Dog/Disc Dog

Frisbee Dog is a concept from the USA. It sounds simple (you throw a frisbee, your dog catches it), but is actually quite complex as it is a choreography of various elements that have to be performed as accurately as possible. There are, however, different varieties, and also competitions: if you don't want to follow a specific pattern, you can also play 'Freestyle' by spontaneously adding different elements together.

Game ideas for the deaf dog

Who needs his ears when he can smell?

So your dog can't hear – but that's not a problem! He depends far more on his fine nose anyway, and just loves to snoop and sniff around. Not only hunting dogs like to go in search of clues; a little scent-tracking game in your garden – or on the road during a walk – is a great idea, and, in bad weather, his nose can easily be kept busy in the living room.

Instead of simply supplying his bowl of food as usual, every now and again, ask him to work for it! Take a morsel of his food and hide it in a couple of locations (quite close to him initially, and gradually increase the distance). His nose will inform him immediately that there is something delicious nearby, and the search begins. By the way, you can teach him a hand signal for this. As he sniffs around, now and then looking at you, signal him with the chosen sign that he should continue searching (we clap our hands and then hold them out as in an embrace). Repeat the signal if your dog finds one portion of food to encourage him to continue searching. Eventually, he will realise that your strange signal means there is still something good for him to find.

Make the search game harder from time to time, so that it doesn't get boring; hide his food in tall grass, among pebbles, or in an old washing up bowl, for example. Because he will observe you carefully, make sure your deaf dog doesn't see where you hide the food. Yes, even dogs cheat! Distract him with a toy, or hide the food in another room.

Jack loves searching for his favourite toy in the garden – he can't get enough of this game. When we got Chocolate, she would watch where we hid Jack's toy and run off with it – spoilsport! See *Smellorama – nose games for dogs* (Hubble and Hattie) for some great game ideas.

Scenting games provide mental stimulation – and are great fun, too!

Games for clever dogs

Your dog may play the fool at

times, but he is certainly not stupid! Scientists now agree that many dogs are more intelligent than dolphins, and even monkeys. Therefore, your companions should be challenged not only physically but mentally, too. Give him some tricky tasks and tasty rewards every day; it's easy to come up with suitable games.

For example, many objects used for intelligence games can be found in your home. A few empty yogurt pots upturned on the floor are the cue for your dog to sniff out which one has a tasty treat hidden under it. Or you could try a more complicated game: place a favourite toy and a bowl in front of him, and hide a reward under a cup. Before your dog gets his treat, he must first put the toy in the bowl.

If you are good at DIY, you could make something out of wood which will last a bit longer, and uses the same principle. Get a plank of wood and make a few recesses in which to hide the treats. However, so your dog cannot reach them too easily, cover the treats with cones or pieces of wood the same size as the recesses in the plank so that they fit neatly over the top.

If you're not keen on the idea of DIY, many pet shops – in the High Street and online – sell intelligence games for dogs.

There are loads of different intelligence games you could play, and if you need some inspiration, take a look at *Dog Games – stimulating play to entertain your dog and you*, published by Hubble and Hattie, search the internet, or ask your friends which games they play.

Chapter 8
Case histories

Of course, it's all very well researching the theory about life with a deaf dog, but, in practice, it might appear completely different, so I have included case histories from owners of deaf dogs. I hope these success stories encourage you to embark on the unique adventure that is life with a deaf dog!

Tamara Kerbler and her bilaterally deaf French Bulldog, Chilly

Chilly was born on March 20 in 2004. At the time I got her, I didn't know she was deaf. She was eight weeks old and it was love at first sight.

After just one week, we already suspected that something was wrong and took her to the vet, who confirmed to us that Chilly was deaf in both ears. I had no idea how to train a deaf dog, or what life would be like with her, but, with patience and lots of goodies, Chilly learned really fast that there are rules in life that she had to comply with.

Now that she maintains eye contact, it's possible for her to run around without a lead on, but as soon as I hold the lead up in the air she immediately runs over to me and sits. We didn't even have to teach her this.

continued over

For me, it was always important to be totally involved with my dog and to watch her reactions to my training. I repeated hand signals again and again until she learnt what they meant.

Chilly is very willing to learn. She fixes her eyes on the face of a human being. It's always fun to watch her when other people talk and accompany their words with gestures. Chilly watches our body language and tilts her head with her eyes fixed on us. Communication with a deaf dog is so much easier if you learn all the hand signals, as these will expand their 'vocabulary' considerably.

Essentially, Chilly is not any different to other dogs, although she barks a little less, and doesn't join in when other dogs bark. If she decides she doesn't want to 'hear' then she simply looks the other way – she worked this one out very quickly! For her, everything is a game.

Chilly has mastered lots of different commands, including: sit! (thumbs up), stay! (palm down), walk! (pointing the index finger down), go to your bed! (a hand in the direction that you want her to go), lie on your back! (hold your hand like a gun), spinning top! (make a circular motion with your thumb, index finger and middle finger), and bark! (index and middle fingers pointing to your mouth). The most important sign for me is that which tells her I love her, and that she has done something good (moving the forefinger and little finger back and forth). If she is being naughty, I point my finger and stamp my foot on the floor, and she lies on her back immediately. It has been my experience that not only do these hand signals work very well, but that Chilly obeys them better than a dog who can hear, because she's not distracted by any noise.

Chilly gets on well with everything and everyone, whether big or small dogs, and even guinea pigs and hamsters. For her, everything is fun and exciting. The only thing that she still doesn't understand to this day is retrieving, which, when asked to do, she is always distracted by something else. And if that something smells particularly good, it is eaten at once.

She is a happy dog who loves to play the fool sometimes, especially when playing around with other French Bulldogs.

When Chilly came to us, we had a second dog, who helped her

to settle in quickly. In my opinion it is very helpful to have a second dog in the house when it comes to deaf dogs. Meanwhile, we are a top team. If she's unhappy, I notice immediately, and vice versa. Chilly always notices when I don't feel well.

In the beginning, one of my greatest fears was that she might snap out of fear because of her disability. We have precluded this by waking her up or touching her at the most unlikely time, then rewarding her by stroking. This worked wonderfully for us. If I scare her, she gets straight on her back! Chilly has never bitten anyone and is very gentle with children.

She also loves riding in the car; in fact, she loves going everywhere with me – even if we are separated only briefly, she always shows tremendous joy at our reunion. Chilly is also very good for a few hours alone at home. I work until noon, which is no problem because she sleeps through half the day anyway. Sometimes I even have to wake her up when I get home!

Of course, there have been moments when I have wished that Chilly could hear; for example, before she had an anaesthetic at the vets. I felt totally helpless and would like to have talked to her gently beforehand. But I'm convinced we have a kind of telepathy.

In summary, I have to say that living with her is just wonderful and I wouldn't swap her for the world. The relationship between us is very special. Everyone who knows Chilli think she is lovely and wants to steal her from me! For me, there is nothing better than my little dog and if I was to get another, I would choose a deaf dog every time. Chilly is the best thing that ever happened to me.

Simona Peters tells the story from the perspective of her bilaterally deaf dog, Luna, a Husky mix

In the spring of 2002, my family very unkindly decided they didn't want me anymore, and gave me to a shelter. Soon, my carer realised that there was something wrong with me. I was friendly to all humans and dogs, loved to run around, fetch sticks and play, but when I was called, I didn't respond.

At first, everyone thought that I had not had any training – and at almost a year old. But, although I'm very greedy, when I was called for dinner, I didn't come running. It soon became clear to all that I AM DEAF!

The shelter staff decided to use sign language with me, which was not easy because spoken language is very important for people. What signs should they use? What rules should be observed? How do you teach all that to a deaf dog? Happily the solution wasn't too hard because I am only deaf – not stupid.

My carer began to wave to me if she wanted me to go to her. Of course, I made sure I got it right, because I was always praised, often with a piece of sausage. I quickly learned the signs for 'sit!' and 'stay!' although the signals had to be very clear. If people waved their hands about wildly, I reacted very nervously and presented all my tricks to them in the hope I might get something right.

Despite trying really hard, I stayed in the shelter, because, after all, who wants a deaf dog?

In June 2002, though, I got lucky when my new owners came to see me. After several visits and walks with my mentor, they took me home from the shelter, and now I live in a small blue house with a garden in the country. I share all of this with two cats, but for me that isn't a problem. Sometimes I cuddle up to the boy cat, who I think is in love with me, but the girl cat is mean – I have to watch out for her quick little paws!

In the first week with my new family, I showed them all my tricks.

I liked it when they let me off the lead and I showed them how fast I could run – I am definitely half-Husky!

With my new family, I go on great walks. I've made many new friends, and discover something new every day. In 2003, our family grew as we have another dog. Scooby is a real coward, terrified of every sound he hears. Of course, I'm not, because I can't hear anything, and when I sleep, I *really* sleep, which means that, on occasion, I am a little frightened when I awake to find that a visitor has arrived and is sitting in the room with us!

Scooby has benefited greatly from my calmness and tranquillity, and I benefit from his body language; through him, I can 'hear' many things that I wasn't previously aware of it. Initially, it was very hard for my mum, as Scooby and I discovered that hunting together is great fun, and we would run off. Now I am a very good girl and always stay close to my mum. Scooby still runs off every now and again, but I tell him off when he comes back.

In 2004 mum and me spent a weekend with the Swiss dog trainer Hans Schlegel. His way of training places no emphasis on oral communication, but on working with the laws of cause and effect and loving treatment. This has not only improved my self-confidence, but mum's, too!

Today, I have mastered 15 hand signals, and live a normal doggie life. Very often, people cannot believe that I'm deaf. I even passed the Companion Dog Certificate with flying colours! So, now you all know that a deaf dog can be a happy and obedient dog, and that this disability need not be a problem!

Anette Otterbach shares her experiences of life with crossbreed Kyra, also deaf in both ears

Life with Kyra, a crossbreed from Greece who has been deaf since birth, is wonderful. Kyra is a very sensitive and very loving dog, who especially loves children. She knows if a child is mentally retarded, and very careful snuggles up to them, almost as if she is trying to tell them, "I like you just the way you are." She has a good soul.

We have to pay extra attention to her when she walks without a lead, but she's not too bad. We have a second dog who is very insecure; she suffered badly before she was rescued, and was found in a dumpster in Italy. She relies very much on Kyra, when you might expect it would be the other way around.

It's so lovely to see how these dogs blossom and become happy with who they are, secure in the knowledge that they are well-loved, finally.

I can highly recommend life with a deaf dog, even with children in the house. It is also a great learning experience for children to realise that no one and nothing is perfect, yet they learn to love the dog anyway.

Appendix 1
Further reading & websites

Further reading

Hear, Hear! A guide to training a deaf dog by Barry Eaton
ISBN: 9780953303926

Living with a deaf dog by Susan C Becker
ISBN: 9780966005805

Dogs can sign, too: a breakthrough method for teaching your dog to communicate to you by Sean Senechal
ISBN: 9781587613531

Dog Games: stimulating play to entertain your dog and you by Christiane Blenski
Hubble & Hattie
ISBN: 9781845843335

Smellorama: nose games for dogs by Viviane Theby
Hubble & Hattie
ISBN: 9781845842932

Dog Relax – relaxed dogs, relaxed owners by Sabina Pilguj
Hubble & Hattie
ISBN: 9781845843335

Websites

There are some excellent websites to be found on the internet; here are just a few of them:

- www.deaf-dogs-help.co.uk/
- www.dogstrust.org.uk/az/d/deafdogs/default.aspx

- www.wagntrain.com/deaf_dog.htm
- www.dogtrainingclassroom.com/deaf-dog-training.html
- www.dfordog.co.uk/deafdogs_more.htm
- www.deafdal.co.uk/Links.htm
- www.thedeafdognetwork.webs.com/
- www.amorak.net/
- www.agilitynet.co.uk/training/deaf_roundtable.HTML
- www.pawstoadopt.com/
- blindanddeafdogs/

Appendix 2
Veterinary practices with audiometric equipment

Very few veterinary practices offer audiometric testing onsite as it is a specialised practice, and the equipment needed is extremely expensive and requires a lot of storage space.

Your vet should be able to refer you to a canine audiometric specialist, and the following is a list of specialists in the UK, as recommended by the Kennel Club (http://www.thekennelclub.org.uk):

Venues and contacts for hearing testing in the UK
Testing is done by appointment only.

- Small Animal Centre, Animal Health Trust. Lanwades Park, Newmarket, Suffolk CB8 7UU, England.
Contact: Ms Julia Freeman.
Tel: 01638 552 700

- Animal Medical Centre, 511 Willbraham Road, Chorlton-cum-Hardy, Manchester M21 0UB, England.
Contact: Mr Pip Boydell.
Tel: 0161 881 3329

- Hearing Assessment Clinic (Mobile), Seadown Veterinary Hospital, 1 Frost Lane, Hythe, Hants, England.
Contact: K Morris MRCVS.
Tel: 02380 842237

- Small Animals Clinic, Royal Dick Veterinary College, Edinburgh
Contact: Prof I G Mayhew,
Large Animal Clinic, Department of Veterinary Clinical Studies, Easterbush Veterinary Centre, Easterbush, Roslin, Midlothian, EH25 9RG Scotland.
Tel: 0131 650 1000 is the number for the Vet College switchboard;

ask to be put through to the small animals clinic.

- Church Farm Veterinary Clinic, Neaston Road, Willaston, South Wirral, Liverpool, CH64 2TL, England.
Contact: Mr G Skerritt.
Tel: 0151 327 1885

- Wey Referrals, 125-129 Chertsey Road, Woking, Surrey GU21 5BP, England.
Contact: Ms Sue Fitzmaurice.
Tel: 01483 729194

More great books from Hubble & Hattie

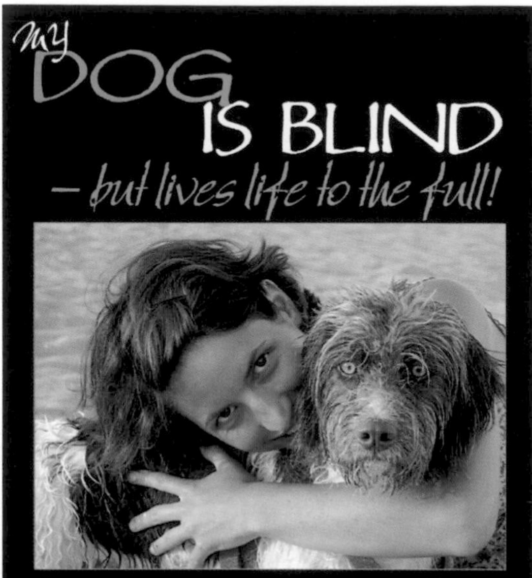

£9.99* each

More details and to order: visit www.hubbleandhattie.com

£9.99* each

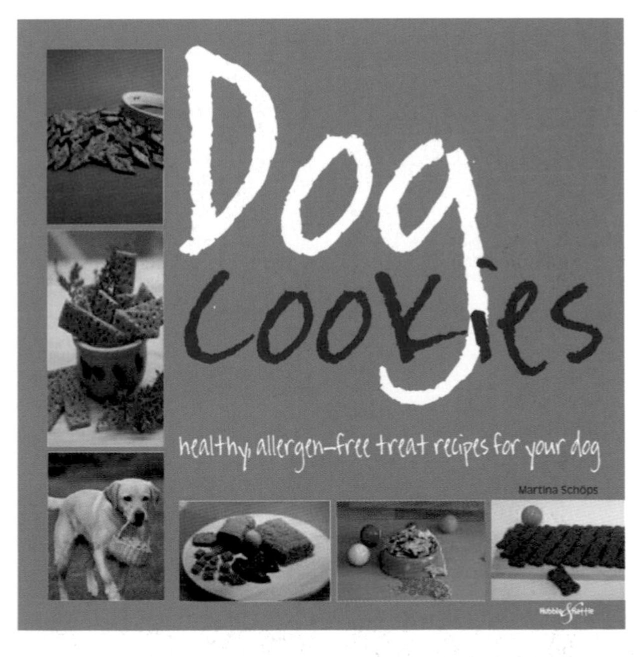

swim to recovery
canine hydrotherapy healing
Hubble & Hattie
Emily Wong
Gentle Dog Care
exercising your puppy
a gentle & natural approach
Hubble & Hattie
Julia Robertson &
Elisabeth Pope
Gentle Dog Care

£12.99* each

living with an older dog
David Alderton & Derek Hall

Hubble & Hattie
Gentle Dog Care

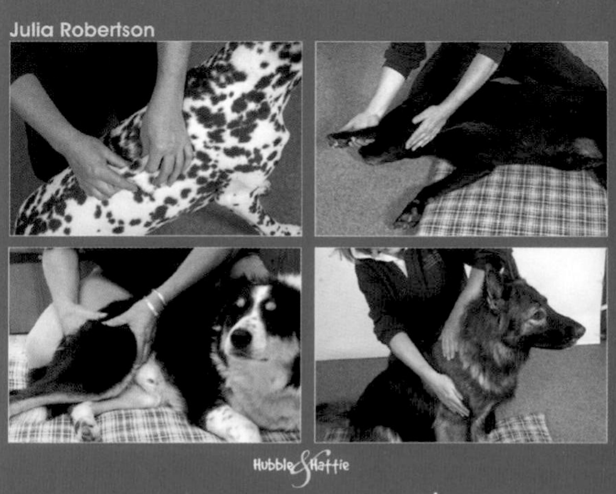
the complete dog massage manual
Julia Robertson
Hubble & Hattie
Gentle Dog Care

dog
SPEAK
recognising and
understanding
behaviour
Hubble & Hattie
cat
SPEAK
Hubble & Hattie
recognising and
understanding
behaviour

£9.99* each

CLEVER
DOG!
LIFE LESSONS FROM THE WORLD'S
MOST SUCCESSFUL ANIMAL
MASTER SALESMAN
MASTER SURVIVOR
MASTER TACTICIAN
MASTER NEGOTIATOR
MASTER LISTENER
MASTER OPPORTUNIST
MASTER OF INTEGRITY
RYAN O'MEARA
Hubble & Hattie
KNOW
YOUR
DOG
The guide
to a
beautiful
relationship
Immanuel
Birmelin
Hubble & Hattie

£9.99*

£14.99*

£14.99*

£19.99*

with my
Dieting DOG
ONE BUSY LIFE, TWO FULL FIGURES
... AND UNCONDITIONAL LOVE
Peggy Frezon
Hubble & Hattie
£9.99*
Dog-friendly gardening
Creating a safe haven for you and your dog
Hubble & Hattie
Karen Bush
£12.99*
walkin' THE DOG
Motorway walks for drivers and dogs
Lezli and David Rees
Hubble & Hattie
£4.99*
walking THE DOG
French motorway walks for drivers and dogs
Lezli Rees
Hubble & Hattie
£4.99*

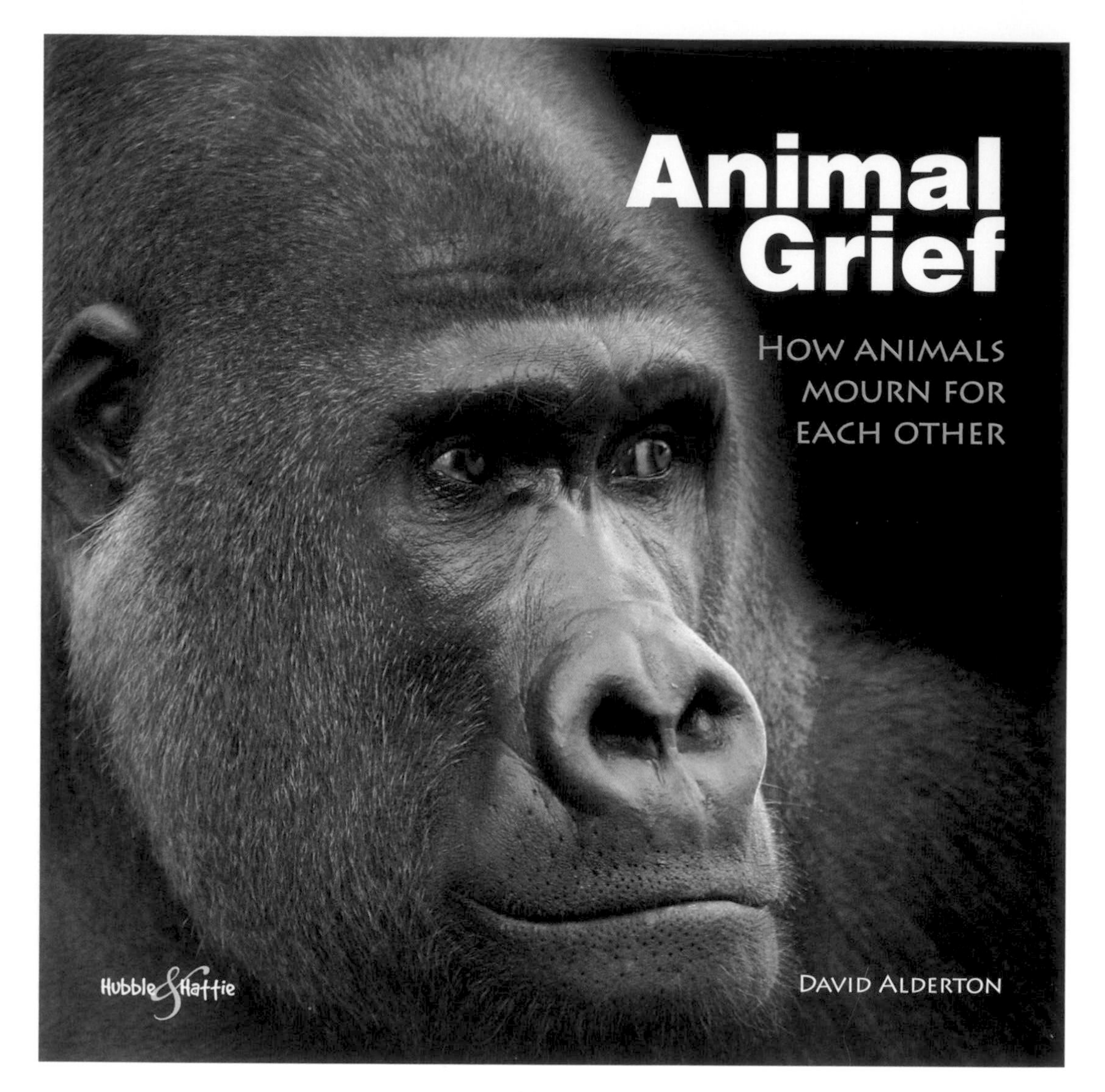

£9.99*

There seems little doubt that animals experience a range of emotions, just as we do; but can they grieve, too ...? This authoritative, rational text is superbly supported by interesting, sensitive photographs, carefully chosen to be reflective of the subject matter.

£12.99*

A critique of traditional dog training and all the myths surrounding it, prompting the reader to look again at why we do certain things with our dogs. It corrects out-of-date theories on alpha status and dominance training, which have been so prominent over the years, and allows you to consider dog training afresh in order to re-evaluate your relationship with your canine companion, ultimately achieving a partnership based on mutual trust, love and respect.

Your dog will thank you for reading this book.

£9.99*

£4.99*

£9.99* each